Ham Radio

How To Get Your Ham Radio License

Table of content

Introduction 3

Chapter 1. Benefits of obtaining an entry level technician class license 5

Chapter 2. Types of Ham Radio Licenses you can obtain 7

Chapter 3. How to learn or study for your first Ham radio license 10

Chapter 4. Components of Ham Radio license exams 13

Chapter 5. Steps for getting your Ham Radio license 22

Conclusion 26

Introduction

Just before you apply for a Ham radio license, there are some things you need to consider, these are in form of questions;

- Why are you interested in in getting an amateur radio and how do you intend or plan to make use of it?
- What is your learning or studying style?
- What do you intend to learn or gain from the licensing procedures.

First of all, when you know your interests and intended use of a Ham radio, you will be rest assured that your expectations and hard-work will be worth the time you devote to obtaining the license. The reason for this question is that it may take you between several weeks or months to prepare for your first Ham radio license. When you eventually pass your exams and obtain your first Ham radio license, you will become so proud of the fact that you have done the right thing by preparing very well to get the license.

You need to ask yourself the question; how do I intend to use my Ham radio license? This also means how you plan to talk to people all around the world or your community, within the band for which you can use your Ham radio. The Ham radio privilege will require a minimum of general class license and that means you need to study and prepare well for the license especially if you hope to expand the radio in the future , and migrate to a superior Ham radio license.

You need to keep in mind that an entry level technician license will place restriction on the privileges you can obtain on voice, and simplex Ham radio operations. An entry level technician license can only give you access to 100 miles of with the use of good system repeaters. Relay stations can easily pick up your transmission, with this setup and then re-broadcast it over a wider area, for a better coverage. If you want to establish contacts between your station and a station outside of your region, you will have to use CW , otherwise known as "Morse Code", however, you will need to upgrade your Ham radio license in order to get this benefit.

Remember, you will not have full access to both digital and non-digital Ham radio band frequencies until you have reached the final and third license category which the Extra class category license. Having a full international connectivity with other radios will also require that you get the final Extra class license but if you just feel like using your Ham Radio locally, then you can have the beginner technician class license that gives you access to the least if privileges.

Chapter 1. Benefits of obtaining an entry level technician class license

Being a beginner Ham radio license, the entry level technician class license come with its own restrictions, just like any other category of Ham radio license, but that does not mean, the license does not come with its own array of benefits. With this license, you can engage in conversations with other local Hams, in your area, and even when you operate a mobile Ham radio. Another benefit is that such a license can help you sharpen your communication skills, as you meet new people in the community who has different ideas to share and who wants to make use of your Ham radio to share such ideas.

With an entry level technician class license, you can use your Ham radio to promote or provide public services, such as publicizing local events such as parades, and marathons- there are some local Ham radio operators who make sustainable income from such ventures.

Disaster management is another benefit that can be desired when a Ham Radio becomes a channel where people can get information from. Messages on Ham radio spreads faster than messages relayed on the internet , when it comes to saving lives within a community.

Your main Ham radio will operate on the FM transceiver, and most likely it will be a mobile or hand-held transceiver unit that can operate on either VHF or UHF frequencies. One of the benefits of this is that you can make use of Voice communications at both VHF and UHF frequencies and both. Make sure the general class license is part of your plan in the immediate future, so that you can use your Ham radio's receivers on both VHF and UHF modes for wider coverage.

Chapter 2. Types of Ham Radio Licenses you can obtain

The Amateur Ham radio licenses are regulated by the FCC (Federal Communications Commission) and they were established under the 1934 Federal Communications Commission Act, and this act is subjected to International agreements. The higher the class of your Ham radio license, the more frequencies will be made available to you. Earning each class of license will require different levels of exams- the higher the class of license the tougher the exams. Volunteer Ham radio operators offer these exams, even though they are regulated by the FCC, these volunteers will grade your tests and report them to the FCC before your license applications are considered.

A typical Ham radio license is valid for 10 years, and can be renewed after and you cannot hold more than one license at a time. The restructuring of licensing by the FCC in 2007 ensures that Ham radio operators can quickly move from beginner levels to Amateur levels. There are three major categories of Ham radio licenses;

- Technician license (Amateur class)
- The General license, and
- The amateur extra license (The amateur extra class).

The technician license or amateur license is the entry level license for any Ham radio operator. In order to get this license, you need to register and pass an examination comprising of 35 questions. These questions are based on Radio theories, Radio regulations, and Ham radio operating practices. Once you obtain this license, you will be given all privileges available on amateur radio frequencies (30 MHz and above), and this license will allow you to communicate locally, in addition , you will get some limited privileges on HF(short wave)- this is the band used for some International communications.

The General license will give you some operational privileges on all the amateur radio bands available alongside all its operating modes. The general license will give you the access to worldwide operations, and just like the technician or entry level license, you will be required to pass 35-question examination, but this exam is slightly tougher than the technician license. You must have written and pass the entry level technician exam before you can register for the general license exam.

The amateur extra license is the ultimate Ham radio license you can obtain. It gives you access to all privileges on all available bands and modes. Getting this license is more difficult than other levels of licenses, and it requires that you pass a 50-question examination. You must have passed the other two types of licensing exams before you can sit for the Amateur extra license.

Why do you need to obtain a Ham radio license?

You need to obtain a Ham radio license before you can get on air, this will assure the FCC that you are capable of operating a radio legally within your jurisdiction. Secondly, you need to obtain a Ham radio license because you are responsible for everything that happens within and around your base station, these responsibilities include; ensuring the security of lives and properties or Ham radio equipment.

Chapter 3. How to learn or study for your first Ham radio license

Don't expect to get your Ham radio license on a platter of gold, there are some exams you need to do to get it. You need to ask yourself the question; what is my learning style? There are basically two ways through which you can prepare for your amateur radio license exam; the easiest way is to attend some hours of classes offered by a local Ham radio club, and the second is by self-study. When you attend a local Ham radio club classes, you can perform hands-on experiment as well as a real testing session.

Make sure your expectations will be worth all the hard-work, You need to know how you want to use the Ham radio license; are you hoping to talk to people all around the world, and if this is your aim, you need to note that that there are limited Voice privileges on all amateur radio licenses, except the final Extra category license.

If you find it difficult to study for a Ham radio license on your own, you should consider enrolling at a training class where you will be exposed to the different formats used and most learning classes will end with actual test questions. When you study alone, you may be tempted to memorize answers and these may not make meaning to you, but when you gain the necessary radio theory alongside the operating procedures and then practicalize these things, then you will be able to write down correct answers in license exams.

One of the best possible ways of preparing for your Ham radio exam is to go for some practical sessions in real Ham radio stations. Make sure you ask the Ham radio owner how the equipment were set up, the type of connectors, antenna and cable used, the power source , as well as all tools involved in the set up. You need to ask about the problems encountered when the Ham radio was being setup and how such problems were detected and rectified.

aside from the practical experiences you gain at local ham radio stations, you should also endeavor to get tapes, cassettes and DVD contents of the license exam you are preparing for, this will ensure that you visualize the practical components of your exams, even though there will be no practical exams , but visual memories help more when writing theoretical exams. You need to pay attention to the test or preparation questions while you are reading through, because the answers to such questions can be found when you read from line to line and understand the concepts.

You need to evaluate the most convenient learning style that can help you pass your exam. If you are good in memorizing contents, a selfo-study style may be your best bet, however, if you are not good in memorizing contents, then a combination of a training class with real practical or hands-on experiment , plus the use of DVD study materials can help you assimilate better .

Always keep in mind that Ham radio license exams always come with multiple choice answers, and many of such answers are quite close to one another , and you must study each answer before making your final choice. Make sure you prepare for topic in Radio electronics, procedures and principles, concepts, and theories before you finally register for an exam.

Chapter 4. Components of Ham Radio license exams

The compositions of different classes of Ham radio license can be similar most times, but the level of difficulties of each component increases from one level to the other, and for this reason, you must be familiar with basic and advance terms of operating Ham radio before you register for each exam.

Technician Class License: Level 1

Exam requirement: Technician license exam (35 questions).

Privileges available: 100% VHF/UHF Amateur bands on all frequencies above 30MHz, plus limited operations on some HF bands.

Composition: The Technician class license for Ham radio operations will cover the basic regulations , as well as operating practices of the Ham radio. Your knowledge of the electronics theory and components of equipment used in setting up a Ham radio will also be tested. In order to prepare for this exam, you need to focus on the VHF/ UHF applications. You don't need a Morse code to prepare for this type of license exam. As you know that this license gives you privileges on bands above the 30 MHz , these privileges will include those available on the 2-meter band, which is the most popular bands used by entry level Ham radio operators.

Most technician class license holders make use of their 2-meter hand-held ham radios to communicate with other Ham radios within their location. As a holder of this license, you are entitled to operate FM voice and digital packets (available on computers), you can also operate Television, alongside the single-side band voice plus several other fun modes. With this license, you can also enjoy limited opportunities such as connecting with other radios internationally via satellite connectivity and with the use of some basic ham station equipment.

Holders of technician licenses now have some additional privileges available on some HF frequencies. As a technician with this license, you can now operate on the 80-meter, 40-meter, and 15-meter bands, with the use of CW, alongside the 10-meter band using the digital modes, CW and voice modes.

You may have to check the technician class rule supplement to be aware of any new changes made on this category of Ham radio license, before you register and sit for the exam.

The tech license gives you access to frequencies of 30MHz and higher, and also gives you access to all mediums of communication (including internet or digital). In order to pass the tech license exam, you need to get 26 out of 35 questions correctly. Just before now, Morse code was required to pass this exam, due to the treaties signed Internationally, in order to create a balance between commercial and military radio traffic- these include traffic from Telegrams, ship to ship and ship to shore messages. This arrangement ensure that emergency radio messages are encoded , with the use of the Morse code, but now, the use of the code for individual Ham radio operations has been removed, and that has reduced stress on potential Ham radio licensees.

The General Class License: Level 2

Exam requirement: 35 questions of general exam

Privileges available: All VHF/UHF amateur bands and most privileges available on the HF (from 10 to 160 meters).

Composition: The General class license is the middle-level license in the three-level Ham radio license issued in the United States and some parts of the world. In order for you to upgrade to the General class license, you must hold a Technician class or entry level Ham radio license, and that means you must have passed the technician class examination. Passing a written exam is the only thing you need to do in order to upgrade to the general class license and it provides some extensive privileges on the HF bands- most of these privileges are not available on the entry level technician license.

Once this exam has been completed, all the operating modes alongside most privileges on the amateur 30MHz spectrum and below will be made available for you. The General class license exam is also referred to as "element 3". You can find the valid forms for this exam on the FCC website download or fill it online and send it to the appropriate address.

In order to facilitate the passing of this exam, you can look out for the exam manual which contains the detailed explanation for all 35 questions set.

One they pass the tech class exam, most Ham radio operators start preparing for the General class exam and when you obtain this license, you have reached a huge milestone in Ham radio operation. Just like the Tech license exam, you will have to answer a minimum of 26 questions rightly, from the total 35 questions, in order to obtain this license. The general class license usually covers the same topics covered by the tech class exam, but you need to know more details on such topics.

The Extra Class License: Level 3

The Extra Class License exam also referred to as "Element 4" and it is the highest license of Ham radio operations you can attain. Many people will refer to this license as "Grandfather Class" license.

Exam requirement: 50 questions

License privileges: All privileges available on Amateur bands.

Composition: You can upgrade into the Extra class license category from the General class but you cannot upgrade from entry level technician level into the Extra class until you get the General class. One of the good things about Extra class license exam is that each question has multiple choice answers, and that can help you a lot. There is no Morse code test required for this exam; however, this exam will cover most of the obscure regulations, alongside advance electronics theories, and Ham radio operating practices. You will also be tested on the design of Radio equipment; therefore you need to study this component to pass the exam.

The FCC will award this license only to those who pass the 50-question exam and also hold other categories of license. Keep in mind that the HF band may become overcrowded and that means you must start preparing for the next stage of licensing once you have gained the HF privileges. The Extra Class License should be your ultimate aim if you want to expand your Ham Radio band operations in the nearest future. As an Extra Class Licensee, you can operate unlimitedly, all amateur Ham radio services, and such include the Internet and International radio operations.

You can also make use of the Extra Class Exam manual to prepare for this exam. A typical manual contains exam questions and answer keys from previous and recent years, alongside the detailed FCC exam rules.

You need to understand that the general class license will not access everything you need with your Ham radio; it is the Extra class license that gives you the maximum privilege. The lowest segments of all HF bands are only reserved for the Extra class licensees, and you should know that most expert Morse code operators can be found with this license. If you are interested in contact the rare and most exclusive radios such as DXing which is available on foreign territories or you need to access most foreign territories, then you need to get this license and do all it takes to pass the exam.

In order to pass the Extra class license exam, you need to answer at least 37 of the 50 questions correctly. You will tested on the additional rules and regulations of operating the Ham radio , likewise the more advance electronics will also be asked . More sophisticated operations and advance technical issues, regarding the running and troubleshooting of Ham radio problems will be asked in this exam. Ham radio operators who pass this exam often consider it as a huge achievement.

The Grand-fathered Ham license categories

The FCC has made tremendous changes and progress over the years, as regards the reduction in the number of Ham radio licenses, and that means many license grades have been scrapped and reduced to the 3 categories highlighted above. If you have a Ham license that has been deleted, and you can now renew the deleted categories of licenses. The grand-fathered Ham license for instance has been removed, and 2 of such licenses has been left, these are;

- The Novice license , and
- Advance license

The Novice License was first introduced in 1951 and the exam normally involves 20 questions. This is a 5-word per minute code examination. This exam was administered by an individual who holds a General class license or higher. The Novice category license has a 10-year lifespan and it is renewable, and before now, this license used to be valid for 1 year and the holder would either renew it or take his or her Ham radio off the air. The novice license is still available these days and it is valid for 10 years and it remains renewable just like any other Ham license, but it gives access only to some segments that are restricted on the 3.5 MHz, 7 MHz, 21 MHz, 28 MHz, 222 MHz, and 1296 MHz amateur bands.

The Advance license holder is expected to pass a written exam that has a difficulty higher than the General license exam, but lower than the amateur extra license exam.

The frequency privileges received by holders of advance license are usually between the General and Amateur extra license frequency privileges.

The table below shows the complete privileges for each category of Ham radio license;

License Class	Privileges	Notes
Technician	All privileges for Amateur Ham radio above 50 MHz; limited CW, Phone, and Data privileges below 30 MHz	
General	All Technician privileges plus most amateur HF privileges	
Amateur Extra	All privileges for amateur license	Small exclusive sub-bands are added on 80, 40, 20, and 15 meters.

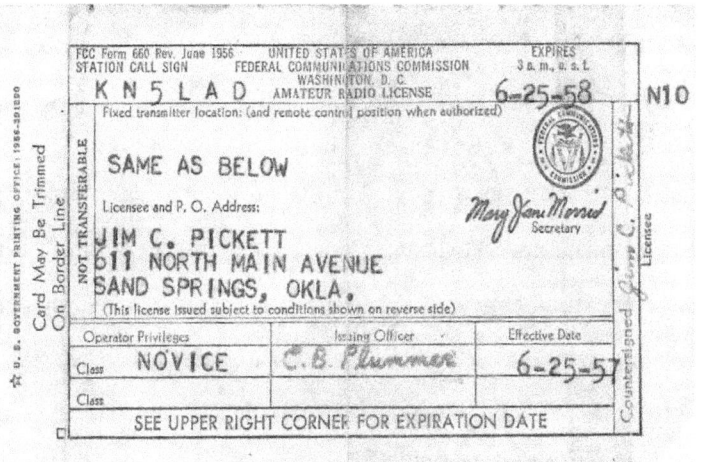

Image: Amateur radio license

Chapter 5. Steps for getting your Ham Radio license

In order to get your Ham Radio License, you need to take certain steps in order to simplify your preparation and ensure that you don't retake the exam. FCC normally set the minimum pass rates for each category of license exam from year to year and without passing the minimum score you will have to retake the exam at a later date. Ham radio license exams can be registered for all through the year, but the FCC sets dates in each quarter of each year.

Step 1: Get a Class

Search and register for a license training class, because this is one of the best possible ways of learning and getting all resources you need to pass your exam. There are a number of online sources you can use to find a Ham radio license coaching class in your community; similarly, you can also search the databases of these sites to get information on upcoming classes. A typical database for checking out upcoming license training classes will provide you an online form where you can fill out your name, city, zip code, state, country and the dates you want to start and complete the class. In most cases the class organizers can also help you register for the license exam.

Once you complete filling the form, you will find the Amateur class closest to you, but you need to keep in mind that not all classes are always available for all categories of exam- the entry level exam training class is usually the most available but vacancies always fill up quickly. While some Amateur training classes are free, you may have to pay a token, and classes normally last between 7-30 days depending on the time schedules.

In order to get the most out of your training class, you need to devote few hours to studying the exam manuals on daily basis and familiarize yourself with how questions are set. Secondly, you need to gather as many resources as possible, to help you upgrade your license within a short period of time. Many amateur license training class operators do provide amateur access to download Amateur Radio question pools, especially from recent and past years. Class training operators also provide all other necessary FCC information that can affect dates and exam modes or changes made to privileges available with different grades of Ham radio licenses.

Step 2: Get a study manual from a local library

You don't have to enroll in a training class just to prepare for your Ham radio license exam. You can study on your own if you can find the study manual at your local library. Some of the popular Ham radio License manuals include the ; ARRL and the Technician class manual by Gordon West. Getting an amateur Ham radio license should be as easy as getting a driver's license, if you follow the orders. In recent changes, the Morse code is no longer required to obtain a Ham radio license, and once you pass your exams and obtain your license, you can enjoy communicating with the public and other Ham radio operators through voice signals, radio signals, digital modes for computers, and Morse codes.

Step 3: Use the study manual to prepare yourself.

Though Ham radio license exams can change in content from year to year however, the mode of setting the questions remains the same. The typical FCC exam has 35 multiple choice questions, except for the Extra class that contains 50 questions. You have less than 2 hours to complete the questions and you should get the outcome of the exam within the same day. Once your exam results have been forwarded to the FCC, you will be contacted to come for your license at a specified date.

Step 4: Use the exam option made available on the ARRL website

The fastest and easiest way to locate an upcoming session of Ham radio license is to use the search option available on the ARRL website. For most locations, you will need to pay a fee of $15 which is the current exam fee, but this will depend on your location, because the exam fees for some locations are completely free. You need to keep in mind that registering for a Ham radio license exam will require that you produce your Social Security Number (SSN).

Step 5: Take your Ham Radio License exam

Once your payment has been confirmed, you will be enrolled in the closest and available exam session. Keep in mind that registering for an exam is based on first-come, and first-serve basis, thus you may be enrolled in the closest upcoming exam session if the current one is filled up. If you don't pass the exam at once season, you can retake it at the upcoming available season, if not the same season.

Step 6: Get your license and find a local radio club

Once you have obtained your Ham radio license, you must search for a local Ham radio club to join. Joining a Ham radio club can be the best possible way to start, especially if you don't have the financial means to start your own radio. Members of your Ham radio club can answer all your questions. Some members of your Ham radio club can lend you a Radio and some equipment or tools to start with.

Conclusion

Having a Ham Radio of your own is a huge accomplishment, and it is worth all the efforts therein. Your Ham radio gives you power within your community, to communicate and be heard. Many Ham radios in the past have grown to become fully licensed radio stations, and the owners are reaping the financial benefits of such. You can use your Ham radio to change your community, and you can expand it in the nearest future and make financial gains from it.

It is quite important that you eliminate all fears, about Ham radio licenses because there is nothing too difficult to do in it. When you understand the procedures, concepts and regulations behind the equipment and operations of a Ham radio, you will find it much easier to pass the license exam , and you will even find it encouraging enough to aim for higher categories of licenses in order to enjoy the full benefits of operating on lower bands.

www.ingramcontent.com/pod-product-compliance
Lightning Source LLC
Chambersburg PA
CBHW050038230526
45470CB00003B/1336